Summary

Navigating the Internet requires using addresses and corresponding names that identify the location of individual computers. The Domain Name System (DNS) is the distributed set of databases residing in computers around the world that contain address numbers mapped to corresponding domain names, making it possible to send and receive messages and to access information from computers anywhere on the Internet. Many of the technical, operational, and management decisions regarding the DNS can have significant impacts on Internet-related policy issues such as intellectual property, privacy, Internet freedom, e-commerce, and cybersecurity.

The DNS is managed and operated by a not-for-profit public benefit corporation called the Internet Corporation for Assigned Names and Numbers (ICANN). Because the Internet evolved from a network infrastructure created by the Department of Defense, the U.S. government originally owned and operated (primarily through private contractors) the key components of network architecture that enable the domain name system to function. A 1998 Memorandum of Understanding (MOU) between ICANN and the Department of Commerce (DOC) initiated a process intended to transition technical DNS coordination and management functions to a private-sector not-for-profit entity. Additionally, a contract between DOC and ICANN authorizes ICANN to perform various technical functions such as allocating IP address blocks, editing the root zone file, and coordinating the assignment of unique protocol numbers. By virtue of this contract and two other legal agreements, DOC exerts a legacy authority and stewardship over ICANN, and arguably has more influence over ICANN and the DNS than other national governments.

On March 14, 2014, the DOC's National Telecommunications and Information Administration (NTIA) announced its intention to transition its stewardship role and procedural authority over key domain name functions to the global Internet multistakeholder community. If a satisfactory transition and Internet governance mechanism can be achieved, NTIA will let its contract with ICANN expire on September 30, 2015. NTIA has stated that it will not accept any transition proposal that would replace the NTIA role with a government-led or an intergovernmental organization solution.

The 113th Congress is likely to closely examine the benefits and risks of NTIA's proposed transition of its authority over ICANN. As a transition plan is developed by ICANN and the Internet community, Congress will likely monitor and evaluate that plan, and seek assurances that an Internet and domain name system free of U.S. government stewardship will remain stable, secure, resilient, and open. Congress will also likely continue to monitor ICANN's rollout of the new generic top level domain (gTLD) program, while also assessing to what extent ongoing and future intergovernmental telecommunications conferences constitute an opportunity for some nations to increase intergovernmental control over the Internet. How these and other DNS-related issues (such as intellectual property, cybersecurity, and privacy) are ultimately addressed and resolved could have profound impacts on the continuing evolution of ICANN, the DNS, and the Internet.

Meanwhile, H.R. 4342 (the DOTCOM Act) was approved by the House Energy and Commerce Committee on May 8, 2014, to prohibit the NTIA from relinquishing responsibility over the Internet domain name system until the Government Accountability Office (GAO) submits a report to Congress examining the ramifications of the proposed transfer. The language of H.R. 4342 was successfully added as an amendment to H.R. 4435, the FY2015 National Defense Authorization Act, which was passed by the House on May 22, 2014. The Senate's FY2015 National Defense Authorization bill (S. 2410) and the House and Senate FY2015 Commerce,

Justice, Science appropriation bills (H.R. 4660/S. 2437) also address the proposed transition. Additionally, other bills introduced into the 113th Congress (H.R. 4367 and H.R. 4398) would place limits on NTIA's ability to transfer its authority over certain domain name functions.

c111173008

Contents

Figures

Tables

Appendixes

Contacts

iii

Background and History

The Internet is often described as a "network of networks" because it is not a single physical entity but, in fact, hundreds of thousands of interconnected networks linking hundreds of millions of computers around the world. Computers connected to the Internet are identified by a unique Internet Protocol (IP) number that designates their specific location, thereby making it possible to send and receive messages and to access information from computers anywhere on the Internet. Domain names were created to provide users with a simple location name, rather than requiring them to use a long list of numbers. Top Level Domains (TLDs) appear at the end of an address and are either a given country code, such as .jp or .uk, or are generic designations (gTLDs), such as .com, .org, .net, .edu, or .gov. The Domain Name System (DNS) is the distributed set of databases residing in computers around the world that contain the address numbers, mapped to corresponding domain names. Those computers, called root servers, must be coordinated to ensure connectivity across the Internet.

The Internet originated with research funding provided by the Department of Defense Advanced Research Projects Agency (DARPA) to establish a military network. As its use expanded, a civilian segment evolved with support from the National Science Foundation (NSF) and other science agencies. While there were (and are) no formal statutory authorities or international agreements governing the management and operation of the Internet and the DNS, several entities played key roles in the DNS. For example, the Internet Assigned Numbers Authority (IANA), which was operated at the Information Sciences Institute/University of Southern California under contract with the Department of Defense, made technical decisions concerning root servers, determined qualifications for applicants to manage country code TLDs, assigned unique protocol parameters, and managed the IP address space, including delegating blocks of addresses to registries around the world to assign to users in their geographic area.

NSF was responsible for registration of nonmilitary domain names, and in 1992 put out a solicitation for managing network services, including domain name registration. In 1993, NSF signed a five-year cooperative agreement with a consortium of companies called InterNic. Under this agreement, Network Solutions Inc. (NSI), a Herndon, VA, engineering and management consulting firm, became the sole Internet domain name registration service for registering the .com, .net., and .org. gTLDs.

After the imposition of registration fees in 1995, criticism of NSI's sole control over registration of the gTLDs grew. In addition, there was an increase in trademark disputes arising out of the enormous growth of registrations in the .com domain. There also was concern that the role played by IANA lacked a legal foundation and required more permanence to ensure the stability of the Internet and the domain name system. These concerns prompted actions both in the United States and internationally.

An International Ad Hoc Committee (IAHC), a coalition of individuals representing various constituencies, released a proposal for the administration and management of gTLDs on February 4, 1997. The proposal recommended that seven new gTLDs be created and that additional registrars be selected to compete with each other in the granting of registration services for all new second level domain names. To assess whether the IAHC proposal should be supported by the U.S. government, the executive branch created an interagency group to address the domain name issue and assigned lead responsibility to the National Telecommunications and Information Administration (NTIA) of the Department of Commerce (DOC). On June 5, 1998, DOC issued a final statement of policy, "Management of Internet Names and Addresses." Called the White Paper, the statement indicated that the U.S. government was prepared to recognize and enter into agreement with "a new not-for-profit corporation formed by private sector Internet stakeholders to administer policy for the Internet name and address system."[1] In deciding upon an entity with which to enter such an agreement, the U.S. government would assess whether the new system ensured stability, competition, private and bottom-up coordination, and fair representation of the Internet community as a whole.

The White Paper endorsed a process whereby the divergent interests of the Internet community would come together and decide how Internet names and addresses would be managed and administered. Accordingly, Internet constituencies from around the world held a series of meetings during the summer of 1998 to discuss how the New Corporation might be constituted and structured. Meanwhile, IANA, in collaboration with NSI, released a proposed set of bylaws and articles of incorporation. The proposed new corporation was called the Internet Corporation for Assigned Names and Numbers (ICANN). After five iterations, the final version of ICANN's bylaws and articles of incorporation were submitted to the Department of Commerce on October 2, 1998. On November 25, 1998, DOC and ICANN signed an official Memorandum of Understanding (MOU), whereby DOC and ICANN agreed to jointly design, develop, and test the mechanisms, methods, and procedures necessary to transition management responsibility for DNS functions—including IANA—to a private-sector not-for-profit entity.

On September 17, 2003, ICANN and the Department of Commerce agreed to extend their MOU until September 30, 2006. The MOU specified transition tasks which ICANN agreed to address. On June 30, 2005, Michael Gallagher, then-Assistant Secretary of Commerce for Communications and Information and Administrator of NTIA, stated the U.S. government's principles on the Internet's domain name system. Specifically, NTIA stated that the U.S. government intends to preserve the security and stability of the DNS, that the United States would continue to authorize changes or modifications to the root zone, that governments have legitimate interests in the management of their country code top level domains, that ICANN is the appropriate technical manager of the DNS, and that dialogue related to Internet governance should continue in relevant multiple fora.[2]

On September 29, 2006, DOC announced a new Joint Project Agreement (JPA) with ICANN which was intended to continue the transition to the private sector of the coordination of technical functions relating to management of the DNS. The JPA extended through September 30, 2009, and focused on institutionalizing transparency and accountability mechanisms within ICANN. On September 30, 2009, DOC and ICANN announced agreement on an Affirmation of Commitments (AoC) to "institutionalize and memorialize" the technical coordination of the DNS globally and by a private-sector-led organization.[3] The AoC affirms commitments made by DOC and ICANN to ensure accountability and transparency; preserve the security, stability, and resiliency of the DNS; promote competition, consumer trust, and consumer choice; and promote international participation.

---

[1] Management of Internet Names and Addresses, National Telecommunications and Information Administration, Department of Commerce, *Federal Register*, Vol. 63, No. 111, June 10, 1998, 31741.

[2] See http://www.ntia.doc.gov/ntiahome/domainname/USDNSprinciples_06302005.pdf.

[3] Affirmation of Commitments by the United States Department of Commerce and the Internet Corporation for Assigned Names and Numbers, September 30, 2009, available at http://www.ntia.doc.gov/ntiahome/domainname/ Affirmation_of_Commitments_2009.pdf.

ICANN Basics

ICANN is a not-for-profit public benefit corporation headquartered in Los Angeles, CA, and incorporated under the laws of the state of California. ICANN is organized under the California Nonprofit Public Benefit Law for charitable and public purposes, and as such, is subject to legal oversight by the California attorney general. ICANN has been granted tax-exempt status by the federal government and the state of California.[4]

ICANN's organizational structure consists of a Board of Directors (BOD) advised by a network of supporting organizations and advisory committees that represent various Internet constituencies and interests (see **Figure 1**). Policies are developed and issues are researched by these subgroups, who in turn advise the Board of Directors, which is responsible for making all final policy and operational decisions. The Board of Directors consists of 16 international and geographically diverse members, composed of one president, eight members selected by a Nominating Committee, two selected by the Generic Names Supporting Organization, two selected by the Address Supporting Organization, two selected by the Country-Code Names Supporting Organization, and one selected by the At-Large Advisory Committee. Additionally, there are five non-voting liaisons representing other advisory committees.

The explosive growth of the Internet and domain name registration, along with increasing responsibilities in managing and operating the DNS, has led to marked growth of the ICANN budget, from revenues of about $6 million and a staff of 14 in 2000, to revenues of $239 million and a staff of 178 in 2013.[5] ICANN has been traditionally funded primarily through fees paid to ICANN by registrars and registry operators. Registrars are companies (e.g., GoDaddy, Google, Network Solutions) with which consumers register domain names.[6] Registry operators are companies and organizations that operate and administer the master database of all domain names registered in each top level domain (for example VeriSign, Inc. operates .com and .net, Public Interest Registry operates .org, and Neustar, Inc. operates .biz).[7]

Additionally, the collection of fees from the new generic top level domain (gTLD) program could contribute to an unprecedented level of revenue for ICANN in the years to come. For example, ICANN forecasts revenues of $162 million from the new gTLD application fees in 2013, which is twice the amount of traditional revenues from all other sources.[8]

---

[4] ICANN, *2008 Annual Report*, December 31, 2008, p. 24, available at http://www.icann.org/en/annualreport/annual-report-2008-en.pdf.

[5] ICANN Board Meeting, *FY14 Budget Approval*, August 22, 2013, available at http://www.icann.org/en/about/financials/adopted-opplan-budget-fy14-22aug13-en.pdf.

[6] A list of ICANN-accredited registrars is available at http://www.icann.org/en/registries/agreements.htm.

[7] A list of current agreements between ICANN and registry operators is available at http://www.icann.org/en/registries/agreements.htm.

[8] *FY14 Budget Approval*, p. 4.

c11173008

Figure 1. Organizational Structure of ICANN

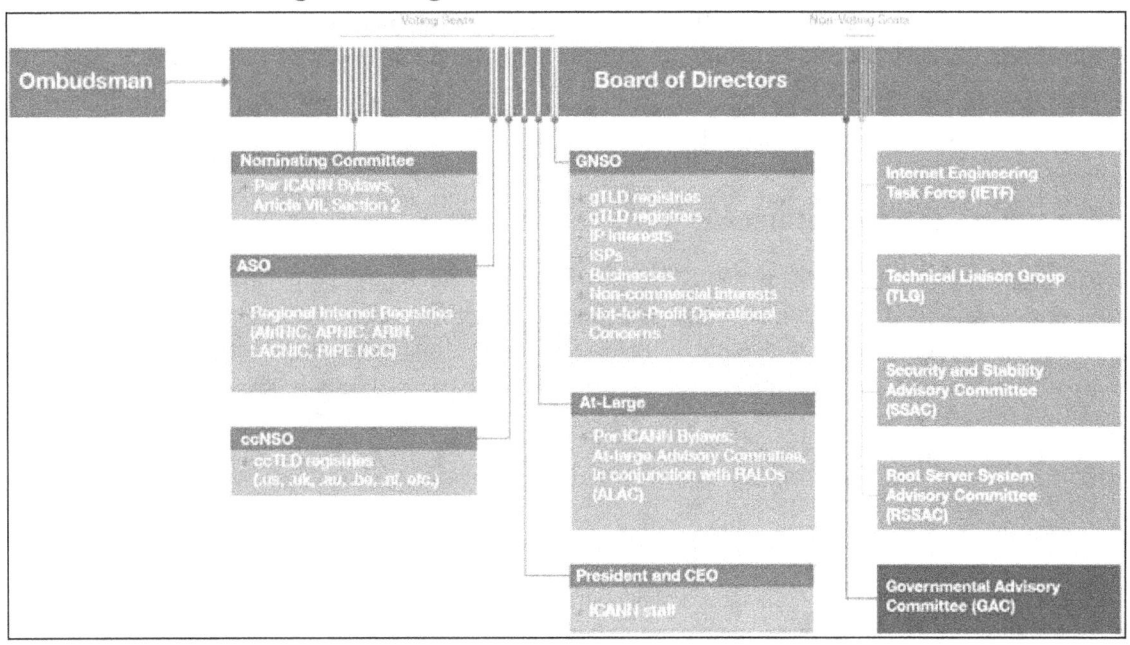

Source: http://www.icann.org/en/groups/chart.

Issues in the 113th Congress

Congressional committees (primarily the Senate Committee on Commerce, Science and Transportation and the House Committee on Energy and Commerce) maintain oversight on how the Department of Commerce manages and oversees ICANN's activities and policies. Other committees, such as the House and Senate Judiciary Committees, maintain an interest in other issues affected by ICANN, such as intellectual property and privacy. The **Appendix** shows a listing of congressional committee hearings related to ICANN and the domain name system dating back to 1997.

ICANN's Relationship with the U.S. Government

The Department of Commerce (DOC) has no statutory authority over ICANN or the DNS. However, because the Internet evolved from a network infrastructure created by the Department of Defense, the U.S. government originally owned and operated (primarily through private contractors such as the University of Southern California, SRI International, and Network Solutions Inc.) the key components of network architecture that enable the domain name system to function. The 1998 Memorandum of Understanding between ICANN and the Department of Commerce initiated a process intended to transition technical DNS coordination and management functions to a private-sector not-for-profit entity. While the DOC plays no role in the internal governance or day-to-day operations of ICANN, the U.S. government, through the DOC, retains a role with respect to the DNS via three separate contractual agreements. These are

- the Affirmation of Commitments (AoC) between DOC and ICANN, which was signed on September 30, 2009;

- the contract between IANA/ICANN and DOC to perform various technical functions such as allocating IP address blocks, editing the root zone file, and coordinating the assignment of unique protocol numbers; and

- the cooperative agreement between DOC and VeriSign to manage and maintain the official DNS root zone file.

Affirmation of Commitments

On September 30, 2009, DOC and ICANN announced agreement on an Affirmation of Commitments (AoC) to "institutionalize and memorialize" the technical coordination of the DNS globally and by a private-sector-led organization.[9] The AoC succeeds the concluded Joint Project Agreement (which in turn succeeded the Memorandum of Understanding between DOC and ICANN). The AoC has no expiration date and would conclude only if one of the two parties decided to terminate the agreement.

Buildup to the AoC

Various Internet stakeholders disagreed as to whether DOC should maintain control over ICANN after the impending JPA expiration on September 30, 2009. Many U.S. industry and public interest groups argued that ICANN was not yet sufficiently transparent and accountable, that U.S. government oversight and authority (e.g., DOC acting as a "steward" or "backstop" to ICANN) was necessary to prevent undue control of the DNS by international or foreign governmental bodies, and that continued DOC oversight was needed until full privatization is warranted. On the other hand, many international entities and groups from countries outside the United States argued that ICANN had sufficiently met conditions for privatization, and that continued U.S. government control over an international organization was not appropriate. In the 110th Congress, Senator Snowe introduced S.Res. 564, which stated the sense of the Senate that although ICANN had made progress in achieving the goals of accountability and transparency as directed by the JPA, more progress was needed.[10]

On April 24, 2009, NTIA issued a Notice of Inquiry (NOI) seeking public comment on the upcoming expiration of the JPA between DOC and ICANN.[11] According to NTIA, a mid-term review showed that while some progress had been made, there remained key areas where further work was required to increase institutional confidence in ICANN. These areas included long-term stability, accountability, responsiveness, continued private-sector leadership, stakeholder participation, increased contract compliance, and enhanced competition. NTIA asked for public comments regarding the progress of transition of the technical coordination and management of the DNS to the private sector, as well as the model of private-sector leadership and bottom-up policy development which ICANN represents. Specifically, the NOI asked whether sufficient

---

[9] Affirmation of Commitments by the United States Department of Commerce and the Internet Corporation for Assigned Names and Numbers, September 30, 2009, available at http://www.ntia.doc.gov/ntiahome/domainname/ Affirmation_of_Commitments_2009.pdf.

[10] In the 110th Congress, S.Res. 564 was referred to the Committee on Commerce, Science, and Transportation. It did not advance to the Senate floor.

[11] Department of Commerce, National Telecommunications and Information Administration, "Assessment of the Transition of the Technical Coordination and Management of the Internet's Domain Name and Addressing System," 74 *Federal Register* 18688, April 24, 2009.

progress had been achieved for the transition to take place by September 30, 2009, and if not, what should be done.

On June 4, 2009, the House Committee on Energy and Commerce, Subcommittee on Communications, Technology, and the Internet, held a hearing examining the expiration of the JPA and other issues. Most members of the committee expressed the view that the JPA (or a similar agreement between DOC and ICANN) should be extended. Subsequently, on August 4, 2009, majority leadership and majority Members of the House Committee on Energy and Commerce sent a letter to the Secretary of Commerce urging that rather than replacing the JPA with additional JPAs, the DOC and ICANN should agree on a "permanent instrument" to "ensure that ICANN remains perpetually accountable to the public and to all of its global stakeholders." According to the committee letter, the instrument should ensure the permanent continuance of the present DOC-ICANN relationship; provide for periodic reviews of ICANN performance; outline steps ICANN will take to maintain and improve its accountability; create a mechanism for implementation of the addition of new gTLDs and internationalized domain names; ensure that ICANN will adopt measures to maintain timely and public access to accurate and complete WHOIS[12] information; and include commitments that ICANN will remain a not-for-profit corporation headquartered in the United States.

Critical Elements of the AoC

Under the AoC, ICANN commits to remain a not-for-profit corporation "headquartered in the United States of America with offices around the world to meet the needs of a global community." According to the AoC, "ICANN is a private organization and nothing in this Affirmation should be construed as control by any one entity."

Specifically, the AoC calls for the establishment of review panels which will periodically make recommendations to the ICANN Board in four areas:

- *Ensuring accountability, transparency and the interests of global Internet users*—the panel will evaluate ICANN governance and assess transparency, accountability, and responsiveness with respect to the public and the global Internet community. The panel will be composed of the chair of ICANN's Governmental Advisory Committee (GAC), the chair of the Board of ICANN, the Assistant Secretary for Communications and Information of the Department of Commerce (i.e., the head of NTIA), representatives of the relevant ICANN Advisory Committees and Supporting Organizations, and independent experts. Composition of the panel will be agreed to jointly by the chair of the GAC and the chair of ICANN.

- *Preserving security, stability, and resiliency*—the panel will review ICANN's plan to enhance the operational stability, reliability, resiliency, security, and global interoperability of the DNS. The panel will be composed of the chair of the GAC, the CEO of ICANN, representatives of the relevant Advisory Committees and Supporting Organizations, and independent experts.

---

[12] Any person or entity who registers a domain name is required to provide contact information (phone number, address, email) which is entered into a public online database (the "WHOIS" database).

Composition of the panel will be agreed to jointly by the chair of the GAC and the CEO of ICANN.

- *Impact of new gTLDs*—starting one year after the introduction of new gTLDs, the panel will periodically examine the extent to which the introduction or expansion of gTLDs promotes competition, consumer trust, and consumer choice. The panel will be composed of the chair of the GAC, the CEO of ICANN, representatives of the relevant Advisory Committees and Supporting Organizations, and independent experts. Composition of the panel will be agreed to jointly by the chair of the GAC and the CEO of ICANN.

- *WHOIS policy*—the panel will review existing WHOIS policy and assess the extent to which that policy is effective and its implementation meets the legitimate needs of law enforcement and promotes consumer trust. The panel will be composed of the chair of the GAC, the CEO of ICANN, representatives of the relevant Advisory Committees and Supporting Organizations, independent experts, representatives of the global law enforcement community, and global privacy experts. Composition of the panel will be agreed to jointly by the chair of the GAC and the CEO of ICANN.

On December 31, 2010, the Accountability and Transparency Review Team (ATRT) released its recommendations to the Board for improving ICANN's transparency and accountability with respect to: Board governance and performance, the role and effectiveness of the GAC and its interaction with the Board, public input and policy development processes, and review mechanisms for Board decisions.[13] At the June 2011 meeting in Singapore, the Board adopted all 27 ATRT recommendations. According to NTIA, "the focus turns to ICANN management and staff, who must take up the challenge of implementing these recommendations as rapidly as possible and in a manner that leads to meaningful and lasting reform."[14] On December 31, 2013, the second ATRT (ATRT2) a follow-up report to the Board with 12 new recommendations (most of which arising from the issues raised in the first ATRT report).[15]

DOC Contract and Cooperative Agreement: IANA and VeriSign

A contract between DOC and ICANN—specifically referred to as the "IANA functions contract"—authorizes ICANN to manage the technical underpinnings of the DNS. Specifically, the contract allows ICANN to perform various critical technical functions such as allocating IP address blocks, editing the root zone file, and coordinating the assignment of unique protocol numbers. Additionally, and intertwined with the IANA functions, a cooperative agreement between DOC and VeriSign (the company that operates the .com and .net registries) authorizes VeriSign to manage and maintain the official root zone file that is contained in the Internet's root servers that underlie the functioning of the DNS.[16]

---

[13] The ATRT final report is available at http://www.icann.org/en/reviews/affirmation/atrt-final-recommendations-31dec10-en.pdf.

[14] NTIA, *Press Release*, "NTIA Commends ICANN Board on Adopting the Recommendations of the Accountability and Transparency Review Team," June 24, 2011, available at http://www.ntia.doc.gov/press/2011/NTIA_Statement_06242011.html.

[15] Available at https://www.icann.org/en/system/files/files/final-recommendations-31dec13-en.pdf.

[16] According to the National Research Council, "The root zone file defines the DNS. For all practical purposes, a top level domain (and, therefore, all of its lower-level domains) is in the DNS if and only if it is listed in the root zone file. (continued...)

c11173008

By virtue of these legal agreements, the DOC has authority over the root zone file, meaning that the U.S. government can approve or deny changes or modifications made to the root zone file (changes, for example, such as adding a new top level domain). The June 30, 2005, U.S. government principles on the Internet's domain name system stated the intention to "preserve the security and stability" of the DNS, and asserted that "the United States is committed to taking no action that would have the potential to adversely impact the effective and efficient operation of the DNS and will therefore maintain its historic role in authorizing changes or modifications to the authoritative root zone file."[17]

The JPA was separate and distinct from the DOC legal agreements with ICANN and VeriSign. As such, the expiration of the JPA and the establishment of the AoC did not directly affect U.S. government authority over the DNS root zone file. Foreign governmental bodies have long argued that it is inappropriate for the U.S. government to maintain that exclusive authority over the DNS.

On July 2, 2012, NTIA announced the award of the most recent (and current) IANA functions contract to ICANN through September 30, 2015 (with an option to extend the contract through September 2019). The contract includes a separation between the policy development of IANA services and the implementation by the IANA functions contractor. The contract also features "a robust company-wide conflict of interest policy; a heightened respect for local national law; and a series of consultation and reporting requirements to increase transparency and accountability."[18] The IANA contract continued to specify that the contractor must be a wholly U.S. owned and operated firm or a U.S. university or college; that all primary operations and systems shall remain within the United States; and that the U.S. government reserves the right to inspect the premises, systems, and processes of all facilities and components used for the performance of the contract.

NTIA Intent to Transition Stewardship of the DNS

The IANA functions contract with ICANN and the cooperative agreement with Verisign give NTIA the authority to maintain a stewardship and oversight role with respect to ICANN and the domain name system. On March 14, 2014, NTIA announced its intention to transition its stewardship role and procedural authority over key domain name functions to the global Internet multistakeholder community.[19] If a satisfactory transition can be achieved, NTIA will let its IANA functions contract with ICANN expire on September 30, 2015.

As a first step, NTIA is asking ICANN to convene interested global Internet stakeholders (both from the private sector and governments) to develop a proposal to achieve the transition.

(...continued)

Therefore, presence in the root determines which DNS domains are available on the Internet." See National Research Council, Committee on Internet Navigation and the Domain Name System, *Technical Alternatives and Policy Implications, Signposts on Cyberspace: The Domain Name System and Internet Navigation*, National Academy Press, Washington, DC, 2005, p. 97.

[17] See http://www.ntia.doc.gov/ntiahome/domainname/USDNSprinciples_06302005.pdf.

[18] NTIA, Press Release, "Commerce Department Awards Contract for Management of Key Internet Functions to ICANN," July 2, 2012, available at http://www.ntia.doc.gov/press-release/2012/commerce-department-awards-contract-management-key-internet-functions-icann.

[19] NTIA, *Press Release*, "NTIA Announced Intent to Transition Key Internet Domain Name Functions," March 14, 2014, available at http://www.ntia.doc.gov/press-release/2014/ntia-announces-intent-transition-key-internet-domain-name-functions.

c11173008

Specifically, NTIA expects ICANN to work collaboratively with parties directly affected by the IANA contract, including the Internet Engineering Task Force (IETF), the Internet Architecture Board (IAB), the Internet Society (ISOC), the Regional Internet Registries (RIRs), top level domain name operators, Verisign, and other interested global stakeholders. In October 2013, many of these groups—specifically, the Internet technical organizations responsible for coordination of the Internet infrastructure—had called for "accelerating the globalization of ICANN and IANA functions, towards an environment in which all stakeholders, including all governments, participate on an equal footing."[20]

NTIA has stated that it will not accept any transition proposal that would replace the NTIA role with a government-led or an intergovernmental organization solution.

In addition, NTIA told ICANN that the transition proposal must have broad community support and address the following four principles:

- support and enhance the multistakeholder model;

- maintain the security, stability, and resilience of the Internet DNS;

- meet the needs and expectation of the global customers and partners of the IANA services; and

- maintain the openness of the Internet.

Supporters of the transition[21] argue that by transferring its remaining authority over ICANN and the DNS to the global Internet community, the U.S. government will bolster its continuing support for the multistakeholder model of Internet governance, and that this will enable the United States to more effectively argue and work against proposals for intergovernmental control over the Internet. Supporters also point out that the U.S. government and Internet stakeholders have, from the inception of ICANN, envisioned that U.S. authority over IANA functions would be temporary, and that the DNS would eventually be completely privatized.[22] According to NTIA, this transition is now possible, given that "ICANN as an organization has matured and taken steps in recent years to improve its accountability and transparency and its technical competence."[23]

Those opposed, skeptical, or highly cautious about the transition[24] point out that NTIA's role has served as a necessary "backstop" which has given Internet stakeholders confidence that the integrity and stability of the DNS is being sufficiently overseen. Critics assert that in the wake of the Edward Snowden NSA revelations, foreign governments might gain more support internationally in their continuing attempts to exert intergovernmental control over the Internet,

---

[20] ICANN, "Montevideo Statement on the Future of Internet Cooperation," October 7, 2013, available at https://www.icann.org/en/news/announcements/announcement-07oct13-en.htm.

[21] ICANN, "Endorsements of the IANA Globalization Process," March 18, 2014, available at https://www.icann.org/en/about/agreements/iana/globalization-endorsements-18mar14-en.pdf.

[22] The Commerce Department's June 10, 1998 Statement of Policy stated that the U.S. government "is committed to a transition that will allow the private sector to take leadership for DNS management." Available at http://www.ntia.doc.gov/legacy/ntiahome/domainname/6_5_98dns.htm.

[23] NTIA, *Press Release*, "NTIA Announced Intent to Transition Key Internet Domain Name Functions," March 14, 2014

[24] See for example: Atkinson, Rob, "U.S. Giving Up Its Internet 'Bodyguard' Role," March 17, 2014, available at http://www.ideaslaboratory.com/2014/03/17/u-s-giving-up-its-internet-bodyguard-role/; and Nagesh, Gauthem, *Wall Street Journal*, "U.S. Plan for Web Faces Credibility Issue," March 18, 2014.

---

and that any added intergovernmental influence over the Internet and the DNS would be that much more detrimental to the interests of the United States if NTIA's authority over ICANN and the DNS were to no longer exist. Another concern regards the development of the transition plan and a new international multistakeholder entity that would provide some level of stewardship over the domain name system. Critics are concerned about the risks of foreign governments—particularly those favoring censorship of the Internet—gaining influence over the DNS through the transition to a new Internet governance mechanism that no longer is subject to U.S. government oversight.

Legislative Activities

On March 27, 2014, Representative Shimkus introduced H.R. 4342, the Domain Openness Through Continued Oversight Matters (DOTCOM) Act. H.R. 4342 would prohibit the NTIA from relinquishing responsibility over the Internet domain name system until GAO submits to Congress a report on the role of the NTIA with respect to such system. The report would include a discussion and analysis of the advantages and disadvantages of the change and address the national security concerns raised by relinquishing U.S. oversight. It would also require GAO to provide a definition of the term "multistakeholder model" as used by NTIA with respect to Internet policymaking and governance. H.R. 4342 was referred to the House Energy and Commerce Committee. On April 2, 2014, the Subcommittee on Communications and Technology held a hearing on the DOTCOM Act.[25] H.R. 4342 was approved by the House Energy and Commerce Committee on May 8, 2014.

On May 22, 2014, the text of the DOTCOM Act was offered by Representative Shimkus as an amendment to H.R. 4435, the National Defense Authorization Act for FY2015. During House consideration of H.R. 4435, the amendment was agreed to by a vote of 245-177. H.R. 4435 was passed by the House on May 22, 2014. The House Armed Services bill report accompanying H.R. 4435 (H.Rept. 113-446) stated the Committee's belief that any new Internet governance structure should include protections for the Department of Defense-controlled .mil generic top level domain and its associated Internet protocol numbers. The Committee also supported maintaining separation between the policymaking and technical operation of root-zone management functions.

On June 2, 2014, the Senate Armed Services Committee reported S. 2410, its version of the FY2015 National Defense Authorization Act. Section 1646 of S. 2410 ("Sense of Congress on the Future of the Internet and the .mil Top-Level Domain") stated that it is the sense of Congress that the Secretary of Defense should "advise the President to transfer the remaining role of the United States Government in the functions of the Internet Assigned Numbers Authority to a global multi-stakeholder community only if the President is confident that the '.MIL' top-level domain and the Internet Protocol address numbers used exclusively by the Department of Defense for national security will remain exclusively used by the Department of Defense." Section 1646 also directed DOD to take "all necessary steps to sustain the successful stewardship and good standing of the Internet root zone servers managed by components of the Department of Defense." In the report accompanying S. 2410 (S.Rept. 113-176), the Committee urged DOD to "seek an agreement through the IANA transition process, or in parallel to it, between the United States and the Internet Corporation for Assigned Names and Numbers and the rest of the global Internet

---

[25] Hearing before the House Energy and Commerce Committee, Subcommittee on Communications and Technology, "Ensuring the Security, Stability, Resilience, and Freedom of the Global Internet," April 2, 2014, available at http://energycommerce.house.gov/hearing/ensuring-security-stability-resilience-and-freedom-global-internet.

stakeholders that the .mil domain will continue to be afforded the same generic top level domain status after the transition that it has always enjoyed, on a par with all other country-specific domains."

On May 8, 2014, the House Appropriations Committee approved H.R. 4660, the FY2015 Commerce, Justice, Science (CJS) Appropriations Act, which appropriates funds for DOC and NTIA. The bill report (H.Rept. 113-448) stated that in order that the transition be more fully considered by Congress, the Committee's recommendation for NTIA does not include any funds to carry out the transition and that the Committee expects that NTIA will maintain the existing no-cost contract with ICANN throughout FY2015. During House consideration of H.R. 4660, an amendment offered by Representative Duffy was adopted on May 30, 2014 (by recorded vote, 229-178) which stated that (section 562) "[n]one of the funds made available by this Act may be used to relinquish the responsibility of the National Telecommunications and Information Administration with respect to Internet domain name system functions, including responsibility with respect to the authoritative root zone file and the Internet Assigned Numbers Authority functions." H.R. 4660 was subsequently passed by the House on May 30, 2014.

On June 5, 2014, the Senate Appropriations Committee reported its version of the FY2015 Commerce, Justice, Science, and Related Agencies Appropriations Act (S. 2437). In the bill report (S.Rept. 113-181) the Committee directed NTIA to conduct a thorough review and analysis of any proposed transition of the IANA contract in order to ensure that ICANN has in place an NTIA approved multi-stakeholder oversight plan that is insulated from foreign government and inter-governmental control. Further, the Committee directed NTIA to report quarterly to the Committee on all aspects of the privatization process and further directed NTIA to inform the Committee, as well as the Committee on Commerce, Science, and Transportation, not less than seven days in advance of any decision with respect to a successor contract. The Committee also expressed its concern that NTIA has not been a strong advocate for U.S. businesses and consumers through its participation in ICANN's Governmental Advisory Committee (GAC), and stated that it awaits "the past due report on NTIA's plans for greater involvement in the GAC and the efforts it is undertaking to protect U.S. consumers, companies, and intellectual property."

Other legislation addressing the proposed transition includes H.R. 4367 (Internet Stewardship Act of 2014, introduced by Representative Mike Kelly on April 2, 2014), which would prohibit NTIA from relinquishing its DNS responsibilities unless permitted by statute; and H.R. 4398 (Global Internet Freedom Act of 2014, introduced by Representative Duffy on April 4, 2014) which would prohibit NTIA from relinquishing its authority over the IANA functions. Both H.R. 4367 and H.R. 4398 were referred to the Committee on Energy and Commerce. Meanwhile, the House Judiciary Committee, Subcommittee on Courts, Intellectual Property, and the Internet, held a hearing on April 10, 2014, that examined the proposed transition.

Multistakeholder Process to Develop a Transition Proposal

ICANN has convened a process through which the multistakeholder community will attempt to come to consensus on a transition proposal. Based on feedback received from the Internet community at its March 2014 meeting in Singapore, ICANN put out for public input and comment a draft proposal of *Principles, Mechanisms and Process to Develop a Proposal to Transition NTIA's Stewardship of the IANA Functions.*[26] Under the draft proposal, a steering

---

[26] Available at http://www.icann.org/en/about/agreements/iana/transition/draft-proposal-08apr14-en.htm.

c11173008

group would be formed "to steward the process in an open, transparent, inclusive, and accountable manner."[27] The steering group would be composed of representatives of each ICANN constituency and of parties directly affected by the transition of IANA functions (for example, Internet standards groups and Internet number resource organizations).

On June 6, 2014, after receiving public comments on the steering group draft proposal, ICANN announced the formation of a Coordination Group which is responsible for preparing a transition proposal.[28] The IANA Stewardship Transition Coordination Group (ICG) is comprised of 30 individuals representing 13 communities.[29] These representatives were selected by their respective communities. On August 27, 2014, the ICG released its charter, which states that its mission is "to coordinate the development of a proposal among the communities affected by the IANA functions."[30] The ICG has also released its Process Timeline, which envisions a series of steps culminating with NTIA approval of the final transition proposal by September 30, 2015.[31]

In parallel with the IANA stewardship transition process, ICANN has also initiated a separate but related process on how to enhance ICANN's accountability. The purpose of this process is to ensure that ICANN will remain accountable to Internet stakeholders if and when ICANN is no longer subject to the IANA contract with the U.S. government. Specifically, the process will examine how ICANN's broader accountability mechanisms should be strengthened to address the potential absence of its historical contractual relationship with the DOC, including looking at strengthening existing accountability mechanisms (e.g., the ICANN bylaws and the Affirmation of Commitments).

The accountability process consists of two Accountability and Governance groups: the Cross Community Group and the Coordination Group.

The *Cross Community Group* has three tasks: to identify issues for discussion or improvement; to appoint participants to the Coordination Group who may be members of the Cross Community Group or from the broader stakeholder groups; and to provide ongoing community input to the Coordination Group. Membership in the Cross Community Group is open to any stakeholder.

The *Coordination Group* is responsible for categorizing and prioritizing issues including those identified by the Cross Community Group; building solution requirements for issues with input from the Cross Community Group; and issuing the final report/recommendations. The Coordination Group will have approximately 21 participants. Ten members will represent the various ICANN stakeholder groups and will be selected by the Cross Community Group. Up to seven advisors to be selected by a Public Experts Group are intended to provide independent advice and research and identify best practices. In addition to providing independent advice, the advisors will work closely with the other members of the Coordination Group in fulfilling all of the Group's work. Other members of the Coordination group include an ICANN staff member, a

---

[27] Ibid.

[28] Details on the Coordination Group are available at https://www.icann.org/resources/pages/process-next-steps-2014-06-06-en#/.

[29] Information on ICG membership is available at https://www.icann.org/resources/pages/icg-members-2014-07-29-en.

[30] Charter for the IANA Stewardship Transition Coordination Group, August 27, 2014, available at https://www.icann.org/en/system/files/files/charter-icg-27aug14-en.pdf.

[31] The IANA Transition Process Timeline is spelled out at https://www.icann.org/en/system/files/files/icg-process-timeline-08sep14-en.pdf.

past participant in the Accountability and Transparency Review Team(s), a liaison with the IANA Stewardship Transition Coordination Group (ICG), and an ICANN Board liaison. On August 19, 2014, ICANN announced the selection of the four-person Public Experts Group,[32] who will select up to seven advisors to sit on the Coordination Group.

Following a public comment period, the Coordination Group will submit its final report to the ICANN Board. The ICANN Board will consider whether to adopt all or parts of it. Any decision by the Board to not implement a recommendation (or a portion of a recommendation) will be accompanied by a detailed rationale.

ICANN, the International Community, and Internet Governance

Because cyberspace and the Internet transcend national boundaries, and because the successful functioning of the DNS relies on participating entities worldwide, ICANN is by definition an international organization. Both the ICANN Board of Directors and the various constituency groups who influence and shape ICANN policy decisions are composed of members from all over the world. Additionally, ICANN's Governmental Advisory Committee (GAC), which is composed of government representatives of nations worldwide, provides advice to the ICANN Board on public policy matters and issues of government concern. Although the ICANN Board is required to consider GAC advice and recommendations, it is not obligated to follow those recommendations.

Many in the international community, including foreign governments, have argued that it is inappropriate for the U.S. government to maintain its legacy authority over ICANN and the DNS, and have suggested that management of the DNS should be accountable to a higher intergovernmental body. The United Nations, at the December 2003 World Summit on the Information Society (WSIS), debated and agreed to study the issue of how to achieve greater international involvement in the governance of the Internet and the domain name system in particular. The study was conducted by the U.N.'s Working Group on Internet Governance (WGIG). On July 14, 2005, the WGIG released its report, stating that no single government should have a preeminent role in relation to international Internet governance. The report called for further internationalization of Internet governance, and proposed the creation of a new global forum for Internet stakeholders. Four possible models were put forth, including two involving the creation of new Internet governance bodies linked to the U.N. Under three of the four models, ICANN would either be supplanted or made accountable to a higher intergovernmental body. The report's conclusions were scheduled to be considered during the second phase of the WSIS held in Tunis in November 2005. U.S. officials stated their opposition to transferring control and administration of the domain name system from ICANN to any international body. Similarly, the 109th Congress expressed its support for maintaining U.S. control over ICANN (H.Con.Res. 268 and S.Res. 323).[33]

The European Union (EU) initially supported the U.S. position. However, during September 2005 preparatory meetings, the EU seemingly shifted its support towards an approach which favored an enhanced international role in governing the Internet. Conflict at the WSIS Tunis Summit over control of the domain name system was averted by the announcement, on November 15, 2005, of

---

[32] https://www.icann.org/news/announcement-2014-08-19-en.

[33] In the 109th Congress, H.Con.Res. 268 was passed unanimously by the House on November 16, 2005. S.Res. 323 was passed in the Senate by Unanimous Consent on November 18, 2005.

an Internet governance agreement between the United States, the EU, and over 100 other nations. Under this agreement, ICANN and the United States maintained their roles with respect to the domain name system. A new international group under the auspices of the U.N. was formed—the Internet Governance Forum (IGF)—which provides an ongoing forum for all stakeholders (both governments and nongovernmental groups) to discuss and debate Internet policy issues. The IGF does not have binding authority and was slated to run through 2010. In December 2010, the U.N. General Assembly renewed the IGF for another five years and tasked the U.N.'s Commission on Science and Technology for Development (CSTD) to develop a report and recommendations on how the IGF might be improved. A Working Group on Improvements to the Internet Governance Forum was formed, which includes 22 governments (including the United States) and the participation of Internet stakeholder groups.

Starting in 2010 and 2011, controversies surrounding the roll-out of new generic top level domains (gTLDs) and the addition of the .xxx TLD led some governments to argue for increased government influence on the ICANN policy development process.[34] Governments such as the United States, Canada, and the European Union, while favoring the current ICANN multistakeholder model of DNS governance, have advocated an enhanced role for the Governmental Advisory Committee (GAC) on ICANN policy decisions. Other nations—such as Brazil, South Africa, and India (referred to as IBSA)—favored the creation of an Internet policy development entity within the U.N. system, whose purview would include integrating and overseeing existing bodies (such as ICANN) that are responsible for the technical and operational functioning of the Internet. A third group of nations, including Russia and China, proposed a voluntary "International Code of Conduct for Information Security," for further discussion in the General Assembly of the U.N. The Code included language that promotes the establishment of a multilateral, transparent, and democratic international management of the Internet.

World Conference on International Telecommunications (WCIT)

The World Conference on International Telecommunications (WCIT) was held in Dubai on December 3-14, 2012. Convened by the International Telecommunications Union (the ITU, an agency within the United Nations), the WCIT was a formal meeting of the world's national governments held in order to revise the International Telecommunications Regulations (ITRs). The ITRs, previously revised in 1988, serve as a global treaty outlining the principles which govern the way international telecommunications traffic is handled.

Because the existing 24-year-old ITRs predated the Internet, one of the key policy questions in the WCIT was how and to what extent the updated ITRs should address Internet traffic and Internet governance. The Administration and Congress took the position that the new ITRs should continue to address only traditional international telecommunications traffic, that a multistakeholder model of Internet governance (such as ICANN) should continue, and that the ITU should not take any action that could extend its jurisdiction or authority over the Internet.

As the WCIT approached, concerns heightened in the 112th Congress that the WCIT might potentially provide a forum leading to an increased level of intergovernmental control over the Internet. On May 31, 2012, the House Committee on Energy and Commerce, Subcommittee on Communications and Technology, held a hearing entitled, "International Proposals to Regulate

---

[34] For more information on this issue, see CRS Report R42351, *Internet Governance and the Domain Name System: Issues for Congress*, by Lennard G. Kruger.

the Internet." To accompany the hearing, H.Con.Res. 127 was introduced by Representative Bono Mack expressing the sense of Congress regarding actions to preserve and advance the multistakeholder governance model. Specifically, H.Con.Res. 127 expressed the sense of Congress that the Administration "should continue working to implement the position of the United States on Internet governance that clearly articulates the consistent and unequivocal policy of the United States to promote a global Internet free from government control and preserve and advance the successful multistakeholder model that governs the Internet today." H.Con.Res. 127 was passed unanimously by the House (414-0) on August 2, 2012.

A similar resolution, S.Con.Res. 50, was introduced into the Senate by Senator Rubio on June 27, 2012, and referred to the Committee on Foreign Relations. The Senate resolution expressed the sense of Congress "that the Secretary of State, in consultation with the Secretary of Commerce, should continue working to implement the position of the United States on Internet governance that clearly articulates the consistent and unequivocal policy of the United States to promote a global Internet free from government control and preserve and advance the successful multistakeholder model that governs the Internet today." S.Con.Res. 50 was passed by the Senate by unanimous consent on September 22, 2012. On December 5, 2012—shortly after the WCIT had begun in Dubai—the House unanimously passed S.Con.Res. 50 by a vote of 397-0.

During the WCIT, a revision to the ITRs was proposed and supported by Russia, China, Saudi Arabia, Algeria, and Sudan that sought to explicitly extend ITR jurisdiction over Internet traffic, infrastructure, and governance. Specifically, the proposal stated that "Member States shall have the sovereign right to establish and implement public policy, including international policy, on matters of Internet governance." The proposal also included an article establishing the right of Member States to manage Internet numbering, naming, addressing, and identification resources.

The proposal was subsequently withdrawn. However, as an intended compromise, the ITU adopted a nonbinding resolution (Resolution 3, attached to the final ITR text) entitled, "To Foster an enabling environment for the greater growth of the Internet." Resolution 3 includes language stating "all governments should have an equal role and responsibility for international Internet governance" and invites Member States to "elaborate on their respective positions on international Internet-related technical, development and public policy issues within the mandate of ITU at various ITU forums.... "

Because of the inclusion of Resolution 3, along with other features of the final ITR text (such as new ITR articles related to spam and cybersecurity), the United States declined to sign the treaty. While the WCIT in Dubai is concluded, the international debate over Internet governance is expected to continue in future intergovernmental telecommunications meetings and conferences. The 113th Congress will likely monitor this ongoing debate and oversee the U.S. government's efforts to oppose any future proposals for intergovernmental control over the Internet and the domain name system. On April 16, 2013, H.R. 1580, a bill "To Affirm the Policy of the United States Regarding Internet Governance," was introduced by Representative Walden. Using language similar to the WCIT-related congressional resolutions passed by the 112th Congress (S.Con.Res. 50 and H.Con.Res. 127), H.R. 1580 states that "It is the policy of the United States to preserve and advance the successful multistakeholder model that governs the Internet." On May 14, 2013, H.R. 1580 was passed unanimously (413-0) by the House of Representatives.

c11173008

Montevideo Statement on the Future of Internet Cooperation

In October 2013, the President of ICANN and the leaders of other major organizations responsible for globally coordinating Internet technical infrastructure[35] met in Montevideo, Uruguay, and released a statement calling for strengthening the current mechanisms for global multistakeholder Internet cooperation. Their recommendations included the following:

- They reinforced the importance of globally coherent Internet operations, and warned against Internet fragmentation at a national level. They expressed strong concern over the undermining of the trust and confidence of Internet users globally due to recent revelations of pervasive monitoring and surveillance.

- They identified the need for ongoing effort to address Internet Governance challenges, and agreed to catalyze community-wide efforts towards the evolution of global multistakeholder Internet cooperation.

- They called for accelerating the globalization of ICANN and IANA functions, towards an environment in which all stakeholders, including all governments, participate on an equal footing.[36]

NETmundial

The day after the Montevideo Statement was released, the President of ICANN met with the President of Brazil, who announced plans to hold an international Internet governance summit in April 2014 that would include representatives from government, industry, civil society, and academia. NETmundial, which was described as a "global multistakeholder meeting on the future of Internet governance," was held on April 23-24, 2014, in Sao Paulo, Brazil.[37] The meeting was open to all interested stakeholders, and was intended to "focus on crafting Internet governance principles and proposing a roadmap for the further evolution of the Internet governance ecosystem."[38]

The outcome of NETmundial produced a nonbinding "NETmundial Multistakeholder Statement"[39] that set forth general Internet governance principles and identified issues to be discussed at future meetings on the future evolution of Internet governance. According to the U.S. government delegation at NETmundial, the meeting outcome reaffirmed the multistakeholder model of Internet governance, endorsed the transition of the U.S. government's stewardship role of IANA functions to the global multistakeholder community, emphasized the importance of strengthening and expanding upon the mandate of the Internet Governance Forum, and underscored the importance of human rights in the implementation of a free and open Internet.[40]

---

[35] The Internet Society, World Wide Web Consortium, Internet Engineering Task Force, Internet Architecture Board, and all five of the regional Internet address registries.

[36] Full statement is available at http://www.icann.org/en/news/announcements/announcement-07oct13-en.htm.

[37] Further information on NETmundial is available at http://netmundial.br/.

[38] Ibid.

[39] Available at http://netmundial.br/wp-content/uploads/2014/04/NETmundial-Multistakeholder-Document.pdf.

[40] United States Diplomatic Mission to Brazil, "Official Statement by the USG Delegation to NETmundial," April 25, 2014, available at http://brazil.usembassy.gov/statementusgdeletationnetmundial.html.

Panel on the Future of Global Internet Cooperation

On November 17, 2013, ICANN announced the formation of a Panel on the Future of Global Internet Cooperation, which will be composed of stakeholders from government, civil society, the private sector, the technical community, and international organizations. Representing a multistakeholder approach to Internet governance, the Panel prepared a report to "include principles for global Internet cooperation, proposed frameworks for such cooperation and a roadmap for future Internet governance challenges."[41] The report, *Towards a Collaborative, Decentralized Internet Governance Ecosystem*, was released in May 2014.[42]

NETmundial Initiative

On August 28, 2014, the creation of a NETmundial Initiative for Internet Governance Cooperation and Development was announced by the World Economic Forum in partnership with ICANN and other governmental, industry, academic, and civil society stakeholders. While having no formal relationship with the April 2014 NETmundial summit held in Brazil, the purpose of the NETmundial Initiative is "to apply the NETmundial Principles to solve issues in concrete ways to enable an effective and distributed approach to Internet cooperation and governance."[43]

Adding New Generic Top Level Domains (gTLDs)

Top Level Domains (TLDs) are the suffixes that appear at the end of an address (after the "dot"). TLDs can be either a country code such as .us, .uk, or .jp, or a generic TLD (gTLD*)* such as .com, .org, or .gov. Prior to ICANN's establishment, there were eight gTLDs (.com, .org, .net, .gov, .mil, .edu, .int, and .arpa). In 2000 and 2004, ICANN held application rounds for a limited number of new gTLDs; there are currently 22 gTLDs in operation. Some are reserved or restricted to particular types of organizations (e.g., .museum, .gov, .travel) and others are open for registration by anyone (.com, .org, .info).[44] Applicants for new gTLDs are typically commercial and non-profit organizations who seek to become ICANN-recognized registries that will establish and operate name servers for their TLD registry, as well as implement a domain name registration process for that particular TLD.

With the growth of the Internet and the accompanying growth in demand for domain names, debate focused on whether and how to further expand the number of gTLDs. Beginning in 2005, ICANN embarked on a long consultative process to develop rules and procedures for introducing and adopting an indefinite number of new gTLDs into the domain name system. A new gTLD can be any word or string of characters that is applied for and approved by ICANN. Between 2008 and 2011, ICANN released seven iterations of its gTLD Applicant Guidebook (essentially the rulebook for how the new gTLD program will be implemented).

---

[41] ICANN, "High-Level Panel Organizes to Address Future of Internet Governance," November 17, 2013, available at https://www.icann.org/en/news/announcements/announcement-2-17nov13-en.htm.

[42] Available at https://www.icann.org/news/announcement-fb-2014-05-20-en.

[43] World Economic Forum, NETmundial Initiative, available at http://www3.weforum.org/docs/WEF_1NetmundialInitiativeBrief.pdf.

[44] The 21 current gTLDs are listed at http://www.iana.org/domains/root/db/#.

On June 20, 2011, the ICANN Board of Directors voted to approve the launch of the new gTLD program, under which potentially hundreds of new gTLDs could ultimately be approved by ICANN and introduced into the DNS. Applications for new gTLDs were to be accepted from January 12 through April 12, 2012, and an application or evaluation fee of $185,000 is required.[45]

ICANN's approval of the new gTLD program has been controversial, with many trademark holders pointing to possible higher costs and greater difficulties in protecting their trademarks across hundreds of new gTLDs. Similarly, governments expressed concern over intellectual property protections, and, along with law enforcement entities, also cited concerns over the added burden of combating various cybercrimes (such as phishing and identity theft) across hundreds of new gTLDs. Throughout ICANN's policy development process, governments, through the Governmental Advisory Committee, advocated for additional intellectual property protections in the new gTLD process. The GAC also argued for more stringent rules that would allow for better law enforcement in the new domain space to better protect consumers. While changes were made, strong opposition from many trademark holders[46] led to opposition from some parts of the U.S. government towards the end of 2011, including the Senate Committee on Commerce, Science and Transportation,[47] the House Committee on Energy and Commerce,[48] the House Judiciary Committee,[49] and the Federal Trade Commission.[50]

At December 2011 House and Senate hearings, ICANN stated its intention to proceed with the gTLD expansion as planned. ICANN defended its gTLD program, arguing that the new gTLDs will offer more protections for consumers and trademark holders than current gTLDs; that new gTLDs will provide needed competition, choice, and innovation to the domain name system; and that critics have already had ample opportunity to contribute input during a seven-year deliberative policy development process.[51] Ultimately, ICANN did not delay the initiation of the new gTLD program, and the application window was opened on January 12, 2012.

On June 13, 2012, ICANN announced it had received 1,930 applications for new gTLDs,[52] including 66 geographic name applications and 116 Internationalized Domain Names (IDNs) in

---

[45] A FAQ for the new gTLD process is available at http://newgtlds.icann.org/applicants/faqs/faqs-en.

[46] The Association of National Advertisers (ANA) has been a leading voice against ICANN's current rollout of the new gTLD program. See ANA webpage, "Say No to ICANN: Generic Top Level Domain Developments," available at http://www.ana.net/content/show/id/icann.

[47] See "Rockefeller Says Internet Domain Expansion Will Hurt Consumers, Businesses, and Non-Profits—Urges Delay," *Press Release*, Senate Committee on Commerce, Science and Transportation, December 28, 2011, available at http://commerce.senate.gov/public/index.cfm?p=PressReleases.

[48] House Committee on Energy and Commerce, "Committee Urges ICANN to Delay Expansion of Generic Top-Level Domain Program," *Press Release*, December 21, 2011, available at http://energycommerce.house.gov/news/PRArticle.aspx?NewsID=9176.

[49] Letter from Representative Goodlatte and Representative Berman to the Secretary of Commerce, December 16, 2011, available at http://www.icann.org/en/correspondence/goodlatte-berman-to-bryson-16dec11-en.pdf.

[50] Letter from FTC to ICANN, December 16, 2011, available at http://www.ftc.gov/os/closings/publicltrs/111216letter-to-icann.pdf.

[51] Testimony of Kurt Pritz, Senior Vice President, ICANN, before the House Committee on Energy and Commerce, Subcommittee on Communications and Technology, December 14, 2011, available at http://republicans.energycommerce.house.gov/Media/file/Hearings/Telecom/121411/Pritz.pdf. The gTLD expansion is also strongly supported by many in the Internet and domain name industry, see letter to Senator Rockefeller and Senator Hutchison at http://news.dot-nxt.com/sites/news.dot-nxt.com/files/gtld-industry-to-congress-gtlds-8dec11.pdf.

[52] A complete list of new gTLD applications is provided at http://newgtlds.icann.org/en/program-status/application-results/strings-1200utc-13jun12-en.

---

scripts such as Chinese, Arabic, and Cyrillic.[53] With the applications received, ICANN moved into the evaluation phase. ICANN will decide whether or not to accept each of the 1,930 new gTLD applications. The process is multi-tiered and complex. Depending on whether an extended evaluation is required, whether there are objections filed requiring dispute resolution, and whether there is string contention (where one or more qualified applicants are applying for the same gTLD), it could take anywhere from 9 to 20 months (from the time the application window closed on May 30) for a new gTLD to be approved and delegated into the domain name system (DNS). All of the rules, procedures, and policies related to the evaluation of the new gTLDs are provided in ICANN's *gTLD Applicant Guidebook, Version 2012-06-04.*[54]

With the first round application period concluded, there remain significant issues in play as the new gTLD program goes forward. First, ICANN has stated that a second and subsequent round will take place, and that changes to the application and evaluation process will be made such that a "systemized manner of applying for gTLDs be developed in the long term."[55]

Second, as the new gTLDs go "live,"[56] many stakeholders are concerned that various forms of domain name abuse (e.g., trademark infringement, consumer fraud, malicious behavior, etc.) could manifest themselves within the hundreds of new gTLD domain spaces. Thus, the effectiveness of ICANN's approach to addressing such issues as intellectual property protection of second level domain names and mitigating unlawful behavior in the domain name space will be of interest as the new gTLD program goes forward.

.xxx and Protecting Children on the Internet

Domain names have been viewed by some policy makers as a tool that could be used to protect children from obscene or indecent material on the Internet. In the 107th Congress, legislation was enacted to create a "kids-friendly top level domain name" that would contain only age-appropriate content. The Dot Kids Implementation and Efficiency Act of 2002 was signed into law on December 4, 2002 (P.L. 107-317), and authorized NTIA to require the .us registry operator (currently NeuStar) to establish, operate, and maintain a second level domain within the .us TLD (kids.us) that is restricted to material suitable for minors.

An opposite approach—establishing an adult content top level domain name that could be filtered by parents—has also been considered. In past Congresses, two bills were introduced to require the Department of Commerce to compel ICANN to establish a mandatory top level domain name (such as .xxx) for material that is deemed "harmful to minors." The bills were S. 2426 (109th Congress), which was introduced by Senator Baucus, and S. 2137 (107th Congress), which was introduced by Senator Landrieu. Neither of those bills advanced beyond introduction.

Meanwhile, as part of its process to add new generic top-level domains (gTLDs), ICANN repeatedly considered (since 2000) whether to allow the establishment of a gTLD for adult content. On June 1, 2005, ICANN announced that it had entered into commercial and technical

---

[53] Application statistics are available at http://newgtlds.icann.org/en/program-status/statistics.

[54] Available at http://newgtlds.icann.org/en/applicants/agb.

[55] *gTLD Applicant Guidebook*, Module 1, p. 1-21.

[56] The first new gTLDs were delegated into the Internet's Root Zone on October 23, 2013. For a listing of delegated new gTLDs, see http://newgtlds.icann.org/en/program-status/delegated-strings.

negotiations with a registry company (ICM Registry) to operate a new ".xxx" domain, which would be designated for use by adult websites. Registration by adult websites into the .xxx domain would be purely voluntary, and those sites would not be required to give up their existing (for the most part, .com) sites.

Announcement of a possible .xxx domain proved highly controversial. With the ICANN Board scheduled to consider final approval of the .xxx domain on August 16, 2005, the Department of Commerce sent a letter to ICANN requesting that adequate additional time be provided to allow ICANN to address the objections of individuals expressing concerns about the impact of pornography on families and children and opposing the creation of a new top level domain devoted to adult content. ICANN's Governmental Advisory Committee (GAC) also requested more time before the final decision. At the March 2006 Board meeting in New Zealand, the ICANN Board authorized ICANN staff to continue negotiations with ICM Registry to address concerns raised by the DOC and the GAC. However, on May 10, 2006, the Board voted 9-5 against accepting the proposed agreement, but did not rule out accepting a revised agreement. Subsequently, on January 5, 2007, ICANN published for public comment a proposed revised agreement with ICM Registry to establish a .xxx domain. However, on March 30, 2007, the ICANN Board voted 9-5 to deny the .xxx domain, citing its reluctance to possibly assume an ongoing management and oversight role with respect to Internet content.[57]

ICM Registry subsequently challenged ICANN's decision before an Independent Review Panel (IRP), claiming that ICANN's rejection of ICM's application for a .xxx gTLD was not consistent with ICANN's Articles of Incorporation and Bylaws. On February 19, 2010, the three-person Independent Review Panel (from the International Centre for Dispute Resolution) ruled primarily in favor of ICM Registry, finding that its application for the .xxx TLD had met the required criteria, and that the ICANN Board's reversal of its initial approval "was not consistent with the application of neutral, objective and fair documented policy."[58]

The IRP decision was not binding; it was the ICANN Board of Directors' decision to determine how to proceed and whether ICM's application to operate a .xxx TLD should ultimately be approved. At ICANN's March 2010 meeting in Nairobi, the Board voted to postpone any decision about the .xxx TLD, and directed ICANN's CEO and general counsel to write a report examining possible options.[59]

On June 25, 2010, at the ICANN meeting in Brussels, the Board voted to allow ICM's .xxx application to move forward. The Board approved next steps for the application, including expedited due diligence by ICANN staff, negotiations between ICANN and ICM on a draft registry agreement, and consultation with ICANN's Governmental Advisory Committee (GAC).

At the December ICANN meeting in Cartegena, Colombia, the ICANN Board passed a resolution stating that while "it intends to enter into a registry agreement with ICM Registry for the .xxx

---

[57] For a discussion of the constitutionality of a .xxx top level domain name, see CRS Report RL33224, *Constitutionality of Requiring Sexually Explicit Material on the Internet to Be Under a Separate Domain Name*, by Henry Cohen.

[58] International Centre for Dispute Resolution, In the Matter of an Independent Review Process: ICM Registry, LLC, Claimant, v. Internet Corporation for Assigned Names and Numbers, Respondent, Declaration of the Independent Review Panel, ICDR Case No. 50 117 T 00224 08, February 19, 2010, p. 70, available at http://safekids.com/documents/irp-panel-declaration-19feb10-en.pdf.

[59] See possible options and public comments at http://icann.org/en/announcements/announcement-2-26mar10-en.htm.

TLD," the Board will enter into a formal consultation with the Governmental Advisory Committee on areas where the Board's decision is in conflict with GAC advice relating to the ICM application.[60]

A February 2011 letter from ICANN to the GAC acknowledged and responded to areas where approving the .xxx registry agreement with ICM would conflict with GAC advice received by ICANN.[61] With the GAC not offering approval of .xxx (and continuing to raise specific objections), the ICANN Board acknowledged that the Board and the GAC were not able to reach a mutually acceptable solution. Ultimately, on March 18, 2011, at the ICANN meeting in San Francisco, the ICANN Board approved a resolution giving the CEO or General Counsel of ICANN the authority to execute the registry agreement with ICM to establish a .xxx TLD. The vote was nine in favor, three opposed, and four abstentions. The .xxx top level domain became available to all registrants starting in December 2011.

ICANN and Cybersecurity

The security and stability of the Internet has always been a preeminent goal of DNS operation and management. One issue of recent concern is an intrinsic vulnerability in the DNS which allows malicious parties to distribute false DNS information. Under this scenario, Internet users could be unknowingly redirected to fraudulent and deceptive websites established to collect passwords and sensitive account information.

A technology called DNS Security Extensions (DNSSEC) has been developed to mitigate those vulnerabilities. DNSSEC assures the validity of transmitted DNS addresses by digitally "signing" DNS data via electronic signature. "Signing the root" (deploying DNSSEC on the root zone) is a necessary first and critical step towards protecting against malicious attacks on the DNS.[62] On October 9, 2009, NTIA issued a Notice of Inquiry (NOI) seeking public comment on the deployment of DNSSEC into the Internet's DNS infrastructure, including the authoritative root zone.[63] On June 3, 2009, NTIA and the National Institute of Standards and Technology (NIST) announced plans to work with ICANN and VeriSign to develop an interim approach for deploying DNSSEC in the root zone.[64] On June 9, 2010, NTIA filed a notice in the *Federal Register* seeking public comments on its testing and evaluation report and its intention to proceed with the final stages of domain name system security extensions implementation in the authoritative root zone.[65] On July 15, 2010, ICANN published the root zone trust anchor and root operators began to serve the signed root zone with actual keys, thereby making the signed root zone available.

---

[60] ICANN, *Adopted Board Resolutions, Cartegena*, December 10, 2010, available at http://www.icann.org/en/minutes/resolutions-10dec10-en.htm#4.

[61] Letter from ICANN to Chair of GAC, February 10, 2011, available at http://icann.org/en/correspondence/jeffrey-to-to-dryden-10feb11-en.pdf.

[62] Internet Corporation for Assigned Names and Numbers, "DNSSEC—What Is It and Why Is It Important?" October 9, 2008, available at http://icann.org/en/announcements/dnssec-qaa-09oct08-en.htm.

[63] Department of Commerce, National Telecommunications and Information Administration, "Enhancing the Security and Stability of the Internet's Domain Name and Addressing System," 73 *Federal Register* 59608, October 9, 2008.

[64] Department of Commerce, National Institute of Standards and Technology, *NIST News Release*, "Commerce Department to Work With ICANN and VeriSign to Enhance the Security and Stability of the Internet's Domain Name and Addressing System," June 3, 2009.

[65] Department of Commerce, National Telecommunications and Information Administration, "Availability of Testing and Evaluation Report and Intent To Proceed With the Final Stages of Domain Name System Security Extensions Implementation in the Authoritative Root Zone," 74 *Federal Register* 32748, June 9, 2010.

c11173008

Ultimately, DNSSEC must be voluntarily adopted by registries, registrars, and the thousands of DNS server operators around the world in order to effectively deploy DNSSEC at all levels to maximize protection against fraudulent DNS redirection of Internet traffic.

Privacy and the WHOIS Database

Any person or entity who registers a domain name is required to provide contact information (phone number, address, email) which is entered into a public online database (the "WHOIS" database). The scope and accessibility of WHOIS database information has been an issue of contention. Privacy advocates have argued that access to such information should be limited, while many businesses, intellectual property interests, law enforcement agencies, and the U.S. government have argued that complete and accurate WHOIS information should continue to be publicly accessible. ICANN has debated this issue through its Generic Names Supporting Organization (GNSO), which is developing policy recommendations on what data should be publicly available through the WHOIS database. On April 12, 2006, the GNSO approved an official "working definition" for the purpose of the public display of WHOIS information. The GNSO supported a narrow technical definition favored by privacy advocates, registries, registrars, and non-commercial user constituencies, rather than a more expansive definition favored by intellectual property interests, business constituencies, Internet service providers, law enforcement agencies, and the Department of Commerce (through its participation in ICANN's Governmental Advisory Committee). At ICANN's June 2006 meeting, opponents of limiting access to WHOIS data continued urging ICANN to reconsider the working definition. On October 31, 2007, the GNSO voted to defer a decision on WHOIS database privacy and recommended more studies. The GNSO also rejected a proposal to allow Internet users the option of listing third party contact information rather than their own private data. Currently, the GNSO is exploring several extensive studies of WHOIS.[66] On June 22, 2011, the ICANN announced the initiation of four separate studies of WHOIS, which were recommended by the Governmental Advisory Committee (GAC) in 2008. The studies examine WHOIS "misuse," WHOIS registrant identification, WHOIS proxy and privacy "abuse," and the feasibility of a WHOIS proxy and privacy reveal study.

Meanwhile, a WHOIS policy review team, established by the Affirmation of Commitments, began its first review of WHOIS policy on October 1, 2010.[67] The team issued its final report on May 11, 2012. The report issued 16 recommendations for strengthening WHOIS, including those related to registrar compliance and improving WHOIS data accuracy and access.[68] On November 8, 2012, the ICANN Board approved a resolution directing the ICANN CEO to launch a new effort to redefine the purpose of collecting, maintaining, and providing access to gTLD registration data, and to consider safeguards for protecting that data.[69] On June 6, 2014, an

---

[66] See ICANN "Whois Services" page, available at http://www.icann.org/topics/whois-services/.

[67] See ICANN "WHOIS Policy Review" page, available at http://www.icann.org/en/reviews/affirmation/review-4-en.htm.

[68] WHOIS Policy Review Team, *Final Report*, May 11, 2012, p. 7-18, available at https://community.icann.org/pages/viewpage.action?pageId=33456480.

[69] ICANN, *Approved Board Resolutions*, "WHOIS Policy Team Report," November 8, 2012, available at http://www.icann.org/en/groups/board/documents/resolutions-08nov12-en.htm.

Expert Working Group released its final report detailing recommendations to the ICANN Board for a next-generation Registration Directory Service to replace the current WHOIS system.[70]

Domain Names and Intellectual Property

Ever since the domain name system has been opened to commercial users, the ownership and registration of domain names has raised intellectual property concerns. The White Paper called upon the World Intellectual Property Organization (WIPO) to develop a set of recommendations for trademark/domain name dispute resolutions, and to submit those recommendations to ICANN. At ICANN's August 1999 meeting in Santiago, the board of directors adopted a dispute resolution policy to be applied uniformly by all ICANN-accredited registrars. Under this policy, registrars receiving complaints will take no action until receiving instructions from the domain-name holder or an order of a court or arbitrator. An exception is made for "abusive registrations" (i.e., cybersquatting and cyberpiracy), whereby a special administrative procedure (conducted largely online by a neutral panel, lasting 45 days or less, and costing about $1,000) will resolve the dispute. Implementation of ICANN's Domain Name Dispute Resolution Policy commenced on December 9, 1999. Meanwhile, the 106th Congress passed the Anticybersquatting Consumer Protection Act (incorporated into P.L. 106-113, the FY2000 Consolidated Appropriations Act). The act gives courts the authority to order the forfeiture, cancellation, and/or transfer of domain names registered in "bad faith" that are identical or similar to trademarks, and provides for statutory civil damages of at least $1,000, but not more than $100,000, per domain name identifier.

Currently, intellectual property is one of the key issues driving the debate over ICANN's addition of new generic top level domain names, with many trademark holders, industry groups, and governments arguing that a proliferation of new gTLDs could compromise intellectual property and increase the costs of protecting trademarks. Domain names have also recently been viewed as a possible way to address piracy of online content. In the 112th Congress, S. 968, the Protecting Real Online Threats to Economic Creativity and Theft of Intellectual Property Act (PROTECT IP), and H.R. 3261, the Stop Online Piracy Act (SOPA), were introduced to prohibit Internet service providers from directing Internet traffic to domain names with infringing content.[71]

Concluding Observations

Many of the technical, operational, and management decisions regarding the DNS can have significant impacts on Internet-related policy issues such as intellectual property, privacy, Internet freedom, e-commerce, and cybersecurity. As such, decisions made by ICANN affect Internet stakeholders around the world. In transferring management of the DNS to the private sector, the key policy question has always been how to best ensure achievement of the White Paper principles: Internet stability and security, competition, private and bottom-up policy making and coordination, and fair representation of the global Internet community. What is the best process to ensure these goals, and how should various stakeholders—companies, institutions, individuals, governments—fit into this process?

---

[70] Available at: https://www.icann.org/en/system/files/files/final-report-06jun14-en.pdf.

[71] See CRS Report R42112, *Online Copyright Infringement and Counterfeiting: Legislation in the 112th Congress*, by Brian T. Yeh.

Controversies such as the new gTLDs and .xxx have led some governments to criticize the ICANN policy making process, and to suggest various ways to increase governmental influence over that process, whether it be an enhanced role for the GAC or a greater role for a U.N.-based or multilateral entity. With the increasing impact of the Internet on virtually all aspects of modern society, some governments argue that they should have an enhanced role in developing Internet policies that will affect their citizens. On the other hand, defenders of the multistakeholder model argue that the phenomenal growth of the Internet has been and will continue to be fostered by a bottom-up, consensus approach, which serves to protect policy decisions from the political and bureaucratic control of national governments and international and multilateral institutions.

Congress is likely to closely examine NTIA's March 14, 2014, proposed transitioning of its authority over ICANN and the DNS to a wholly multistakeholder-driven entity. Congress will likely consider whether the proposed transition is in the best interest of the United States and in the best interest of the Internet. As a transition plan is developed by ICANN and the Internet community, Congress will likely monitor and evaluate that plan, and seek assurances that a DNS free of U.S. government stewardship will remain stable, secure, resilient, and open. As part of its examination, Congress will likely continue assessing to what extent ongoing and future intergovernmental telecommunications conferences constitute an opportunity for some nations to increase intergovernmental control over the Internet, and how effectively NTIA and other government agencies (such as the State Department) are working to counteract that threat. Ultimately, how these issues are addressed could have profound impacts on the continuing evolution of ICANN, the DNS, and the Internet.

Appendix. Congressional Hearings on the Domain Name System

Table A-1. Congressional Hearings on the Domain Name System

Date	Congressional Committee	Topic
April 10, 2014	House Judiciary	"Should the Department of Commerce Relinquish Direct Oversight over ICANN?"
April 2, 2014	House Energy and Commerce	"Ensuring the Security, Stability, Resilience, and Freedom of the Global Internet"
February 5, 2013	House Energy and Commerce	"Fighting for Internet Freedom: Dubai and Beyond"
May 31, 2012	House Energy and Commerce	"International Proposals to Regulate the Internet"
December 14, 2011	House Energy and Commerce	"ICANN"s Top-Level Domain Name Program"
December 8, 2011	Senate Commerce, Science and Transportation	"ICANN's Expansion of Top Level Domains"
May 4, 2011	House Judiciary	"ICANN Generic Top-Level Domains (gTLD) Oversight Hearing"
September 23, 2009	House Judiciary	"Expansion of Top Level Domains and its Effects on Competition"
June 4, 2009	House Energy and Commerce	"Oversight of the Internet Corporation for Assigned Names and Numbers (ICANN)"
September 21, 2006	House Energy and Commerce	"ICANN Internet Governance: Is It Working?"
September 20, 2006	Senate Commerce, Science and Transportation	"Internet Governance: the Future of ICANN"
July 18, 2006	House Financial Services	"ICANN and the WHOIS Database: Providing Access to Protect Consumers from Phishing"
June 7, 2006	House Small Business	"Contracting the Internet: Does ICANN Create a Barrier to Small Business?"
September 30, 2004	Senate Commerce, Science and Transportation	"ICANN Oversight and Security of Internet Root Servers and the Domain Name System (DNS)"
May 6, 2004	House Energy and Commerce	"The 'Dot Kids' Internet Domain: Protecting Children Online"
July 31, 2003	Senate Commerce, Science and Transportation	"Internet Corporation for Assigned Names and Numbers (ICANN)"
September 4, 2003	House Judiciary	"Internet Domain Name Fraud – the U.S. Government's Role in Ensuring Public Access to Accurate WHOIS Data"
September 12, 2002	Senate Commerce, Science and Transportation	"Dot Kids Implementation and Efficiency Act of 2002"
June 12, 2002	Senate Commerce, Science and Transportation	"Hearing on ICANN Governance"

c11173008

Date	Congressional Committee	Topic
May 22, 2002	House Judiciary	"The Accuracy and Integrity of the WHOIS Database"
November 1, 2001	House Energy and Commerce	"Dot Kids Name Act of 2001"
July 12, 2001	House Judiciary	"The Whois Database: Privacy and Intellectual Property Issues"
March 22, 2001	House Judiciary	"ICANN, New gTLDs, and the Protection of Intellectual Property"
February 14, 2001	Senate Commerce, Science and Transportation	"Hearing on ICANN Governance"
February 8, 2001	House Energy and Commerce	"Is ICANN's New Generation of Internet Domain Name Selection Process Thwarting Competition?"
July 28, 1999	House Judiciary	"Internet Domain Names and Intellectual Property Rights"
July 22, 1999	Senate Judiciary	"Cybersquatting and Internet Consumer Protection"
July 22, 1999	House Energy and Commerce	"Domain Name System Privatization: Is ICANN Out of Control?"
October 7, 1998	House Science	"Transferring the Domain Name System to the Private Sector: Private Sector Implementation of the Administration's Internet 'White Paper'"
June 10, 1998	House Commerce	"Electronic Commerce: The Future of the Domain Name System"
March 31, 1998	House Science	"Domain Name System: Where Do We Go From Here?"
February 21, 1998	House Judiciary	"Internet Domain Name Trademark Protection"
November 5, 1997	House Judiciary	"Internet Domain Name Trademark Protection"
September 30, 1997	House Science	"Domain Name System (Part 2)"
September 25, 1997	House Science	"Domain Name System (Part 1)"

Author Contact Information

Lennard G. Kruger
Specialist in Science and Technology Policy
lkruger@crs.loc.gov, 7-7070

www.ingramcontent.com/pod-product-compliance
Lightning Source LLC
Chambersburg PA
CBHW081247170526
45165CB00009B/3231